Making Waves

Albert Einstein: Science & Life

Martin Zarrop

V.

Published in the United Kingdom in 2019
by V. Press,
10 Vernon Grove,
Droitwich,
Worcestershire,
WR9 9LQ.

ISBN: 978-1-9998444-9-3

Cover design © Ruth Stacey, 2019.

Printed in the U. K. by Vernon Print & Design, Droitwich, WR9 8QZ.

Acknowledgments

*A version of Seeking Miss Aether appeared in The Journal (2018). A version of
Entanglement appeared in the collection Moving Pictures (Cinnamon Press, 2016).*

*Heartfelt thanks are due (yet again!) to Cross Border Poets, Postcards From Pluto,
Poem Shed, Real Live Poets and, in particular, Richard Hughes, Robbie Burton, Keith
Lander, Alan Clemo and Vivien Finney for their encouragement and invaluable
feedback.*

V.

Contents

V.

V.

Seeking Miss Aether

"To the woman of my dreams:
I'm a mature, single male
who enjoys the pleasures
of classical physics."

You're out there somewhere,
everywhere. I need you
to make sense of my world.

I accept I'm old-fashioned,
viewpoint unchanging.
It was good enough for Newton.

Young Albert insists
that you're past it
but what does he know.

Forget about relativity,
the expanding universe.
This is bigger.

Darling, I've seen the light
bend, space contort
and I worry.

Where are you, lover?
Invisible siren, sing to me;
there's still time.

Celebrity

Albert Einstein 1879–1955

When I looked in the mirror, I saw him,
that warm smile below a halo of hair,
the intensity in his brown eyes.
It's you, I said
but he shook his head with a *No,*
not me, I'd rather be YOU,
in that strong German accent
I remember from old newsreels.

After that, I became well-known
as an after-dinner speaker
on relativity and gravitation,
reality and the quantum,
philosophy and politics
and how to act disgracefully
with any number of women
who hero-worshipped me.

So much affection, so little time
to decipher the thoughts of God.
In the end they checked my birth certificate,
charged me with Impersonating A Physicist.
The scientists of the world were appalled;
they always claimed I was a mathematician
or, even worse, a kind of philosopher.

"Subtle is the Lord, but malicious He is not.
Nature hides her secret because of her essential loftiness,
but not by means of ruse."

Einstein at Princeton (April 1921)

"What I most admired in [Michele Besso] as a human being is
the fact that he managed to live for many years not only in peace
but also in lasting harmony with a woman - an undertaking
in which I twice failed rather disgracefully."

Einstein to V.Besso (March 1955)

Compass

Munich 1884–94

At the back of the class, a boy smiles.
Closing his eyes, he can visualise
invisible fields of magnetic flux, a plough
gouging curved furrows through iron dust.

A simple mechanism points the way,
a needle quivering in the presence
of something hidden deep, thought-lines
only a flight away from the miraculous.

So much to learn, to explain.
The boy trembles at the mystery.
Patience, tenacity, mathematics –
his private games demand all three

as he constructs a house of cards,
upwards, towards the light.

Johnnie & Dollie

"Oh my! That Johnnie boy!
So crazy with desire,
While thinking of his Dollie,
His pillow catches fire."
Albert to Mileva, Aug 1900

Here in Paradise, it's very beautiful.
We'll climb the Üetliberg, then start on
Helmholtz's electromagnetic theory of light!

Mama has hysterics, takes to her bed,
buries her face in a pillow and cries
as only a Jewish mother can:

You are ruining your future. Like you,
she is a book but you ought to have a wife.
If she gets pregnant...

Her loving son needs this like a 'loch im kopf'.
I don't have a moment's peace in my life;
my parents weep for me as if I had died

but consider this interesting question:
how does electric energy radiate through space
in the case of a sinusoidal alternating current?

Here in Paradise, it is very beautiful.
A thousand wishes and the biggest kisses
from your Johnnie.

Thought Experiments

Patent Office, Bern 1902-03

"I feel in my bones the significance of blood."

Consider this: I stand at a lectern
in Room 86 on the third floor
of a building on the corner
of Speichergasse and Genfergasse.
Flat feet and varicose veins don't
prevent me fulfilling my duty.

Consider a gravel sorter,
an electromechanical typewriter,
the pervasive tick
of the Bern town clock.
In this worldly cloister
I must question everything.

Consider this: the world exists
independent of our senses, but
away from my 'cobbler's trade',
there are other problems.
Why did Papa wish to die alone?
Why did we abandon Lieserl?

Disappointment and guilt –
my senses have no access
to these inner mechanisms.
Here, in Room 86, I hatch
such beautiful thoughts.
Others, I must reject.

Quantum Leap

Photoelectricity, March 1905

The century starts with physics' end.
Newton, Maxwell - they'd said it all:
lumps of matter, waves of light,
fields of force, the apple's fall.

The century starts with another thought:
could light consist of discrete lumps?
Too far, said Planck, *although it works*
but Einstein takes the thought and jumps

to photoelectricity:
electrons ripped from stuff by light,
their energy rising with frequency,
infrared to ultraviolet.

Too far, said Planck, *although it works
and justifies my quantum fudge*
but Einstein proved to be correct,
although the old guard wouldn't budge.

Yes, everything comes in packets now
but, despite his Nobel Prize,
he always remained uncomfortable
with waves of matter, lumps of light.

Unseen

Molecules, April–May 1905

"...all chemists use the atomic theory, [but] a considerable number view it with mistrust, some with positive dislike."
President, London Chemical Society, 1869

He's counting molecules, but they might not exist.
What else can be moving in the liquid depths
of his brain or under Brown's microscope?
What secret currents provoke this endless twitch?

He's counting molecules and, outside the window,
a mindless drunk abandons the safety of a street lamp,
to stagger at random, one constant step at a time
into the darkness. How far will he go?

Einstein looks out, past the light,
his mind on the figure as it fades from sight,
imagining a fatter drunk taking smaller steps
unless the night's warmth quickens uncertain legs.

He turns back to his latest papers, insists
he's finished counting molecules.
They exist.

Torch Bearer
Special Relativity, June 1905

He dances with Galileo below decks,
unaware that the shore is far behind.
Water drips from a tap, goldfish swim
in a bowl, as if they were back in Bern.
The sea is still, remarks Galileo
and they both laugh.

As the sun goes down, they go up top,
face into the wind, greet the old captain
who remarks on *a fine mist on the breeze*.
Einstein switches on his new torch.
It seems to shrink in the moonlight.
I see no mist. He smiles, peers at his watch.

Galileo looks uncomfortable.
We're late for dinner, says Einstein.

An Amusing Thought

September 1905

"Light carries mass with it... The thought is amusing and seductive."
Einstein to Conrad Habicht

...and another thing:
r e l a t i v i t y   i m p l i e s
m a s s   i s   e n e r g y

e n e r g y   e q u a l s
mass times the speed of light squared
an amusing thought

that can't be unthought
mass can be transformed
can God be joking?

nineteen forty-five:
matter lights up a city

...

Equivalence

Bern 1907

Which way is up?
The man in the lift,
its cable severed,
spends his last moments
as an astronaut,
directionless.

Floating or falling?
He ponders
their equivalence
ending up (or down?)
on rubber sheets
of gravitation.

Falling,
the astronaut salutes
and, as reaction
follows action,
performs a slow rotation
before terrestrial matter,
without a single thought,
gets in the way.

Genesis

Gravitation 1907–15

In the beginning, God created the heaven and the earth.
And God said: the laws of physics must be of such a nature
that the things, hereafter called tensors, must apply
to systems of reference in any kind of motion.

And Albert changed paradigms for all the world,
and saw the light in the firmament, that it was curved.
And the whole earth was of one language, of one speech,
with the help of the equation of the geodetic line.

Thus, the heavens and earth were finished, and all the host of them,
generally covariant, and for the planet Mercury,
a rotation of the orbit of 43 seconds of arc per century.
And it was very good, so he rested from all his work.

Objective Reality

Berlin 1914-16

The world won't go away
if we don't observe it, Mitsa.
Our life together has become
impossible, even depressing,
but I can't say why.

There have to be rules:
Do not ask to travel with me.
Reply at once to my questions.
Immediately leave the room
at my request.
Don't ask me to sit with you
or expect any kindness.

I shall be satisfied if the boys
become useful, respected men.
I have great trust in your influence
but you will remain always for me
a severed limb.

My life goes on in full harmony.
I am entirely devoted to reflection,
charmed by the vast horizon,
bothered only when an opaque object
obscures the foreground.

Don't feel sorry for me.
There must be certainty in the world.

The New Violin

Total eclipse, May–November 1919

Empires have fallen
and birds hold their breath
as discs embrace in a darkness
that questions the weight of light.
Principe Island: *Through cloud. Hopeful.*
Sobral, Brazil: *Eclipse splendid.*

The data speak and Moses descends
from the mountain with a paradigm
carved in Riemannian stone.
Only twelve people understand it
and the public has begun to doubt
that two times two still equals four.

The data speak and starlight bends
to the will of strange geometries,
the language of tensor calculus.
Newton is dragged into the light,
fails to compete with the prophesies
of the *Suddenly Famous* Dr Einstein.

At forty, the way ahead is clear,
a new wife walking at his side.
To celebrate, he buys another violin
and glances back, half-expecting to glimpse
Mileva limping a few steps behind.
He sees only admirers.

When You Look at Something

Solvay 1927-now

after Robert Graves

We didn't believe it at first.
Even Planck thought it was
a temporary sticking plaster.
How wrong we were.

Was it a little thing, then?

It was like turning a key, really.
The floodgates opened
and it changed everything,
explained so much.

I hear Albert didn't like it.

Well, that was later when a particle
could also be a wave.
We couldn't say until we looked.
Einstein wanted the world
out there, independent.

That sounds reasonable, doesn't it?

Yes, but at the level of atoms,
when you look at something
you change it.
It can't be helped.

That sounds dodgy – and a bit spooky.

Yes, literally. Most electronics
needs the quantum,
not everyday common sense.
It works, so the attitude is:
'Shut up and calculate'.

Isn't that a bit dangerous?
What about the H-Bomb?

I was coming to that.

"Not Yet Hanged"

Caputh 1929–33

I will not see Caputh again,
or sail my Tümmler on the Havel lakes.

It was paradise, that almost silence,
my fiftieth birthday summer house,
those autumn walks from Waldstrasse 7,
no telephone, just thoughtful words
from cultured visitors and friends
as darkness fell.

The cultured men in brown
stormed in last month, searching
for a dump of Communist arms.
They found a breadknife
and some books they later burned
along with my two cousins.

Five thousand dollars on my head;
what price an individual's dignity?

I trust the League of German Maidens
and the Hitler Youth
will enjoy the Caputh summer house,
once owned by a German man of science,
expropriated from an old Swiss Jew.
I will not see my Germany again.

Einstein on the Beach

Kristallnacht, 10th November 1938

Infinity in a grain of sand
or the craze of a window:
no end to it, this madness.

Einstein walks with Blake
as he weaves warp and weft
through the cosmic fabric.

He has time to tease
secrets out of black holes
from a safe distance

while in the darkness
unknown brothers sweep up
fragments of illusion

before the tsunami.

Dilemma

July 1939–August 1945

I hadn't thought of that at all.

If only he wasn't here
at this point in space-time,

the thud of distant boots
enforcing a new order.

German physics invades
his mind, triggering

fission, chain reaction,
a glimpse of annihilation.

He dreams of another universe
without nationalism, slaughter,

the problem of critical mass.
What's Heisenberg thinking?

I made one great mistake in my life,
Einstein confesses later.

His letter to Roosevelt burns
brighter than the sun,

casts everlasting shadows
on witnesses of stone.

Stammesgenossen

Zionism 1919–48

"One can be internationally-minded without lacking concern for members of one's tribe."

A billion marks for a loaf of bread?
Blame the allies, blame me too,
blame an internationalist,
pacifist, intellectual Jew

or, perhaps, refugees
from the ravaged nations
of the east, fleeing pogroms
with their tribal affiliations.

Hatred was always out there
but these crises have honed
my attitude to the question
of Palestine as home.

If you call it *nationalism*
then *community* is turned
into something ugly.
Let's show what we've learned

through millennia of martyrdom
as the persecuted other,
to live side by side
with the Arab, our brother.

A billion marks for a loaf of bread,
six million of my kinsmen dead.

No

Mileva: August 1948

What did he ever see in me?
A good girl, clever and serious,
small, frail, dark, ugly,
limps a bit but has very nice manners,
talks like a real Novi Sad girl.

Was it the science or Elsa?
No, it doesn't matter now.
My only joy was the children
but I needed a daughter's love.
Lieserl, how I've missed you.

Tete suffers, but a person
does not know how to help him;
the nervous disturbances get worse.
No, the Nobel Prize money
wasn't enough to cover expenses.

It was probably my fault, the fall.
I was visiting Tete in hospital –
my uneven legs, the uneven road
and then a stroke, exhaustion.
No, life is quite senseless.

Even my closest friends admire
Einstein's achievements as if they were
personal virtues, but Helene said:
I no longer care for him.
No, I no longer care for him.

Entanglement

Princeton 1948

A phantom haunts the universe,
a quantum thread that binds our lives
to distant mass, refusing to let go.

Astronomers hold to another truth:
as bodies move apart, attraction fades
and memory weighs nothing out in space.

Shut up and calculate
they tell the homesick astronaut
and yet

I thought I saw her yesterday
and wept.

No Theory of Everything
Unified Field Theories 1917–55

We're held down by the grip of mass
as lightning strikes us from the skies.
What's the name of the one true god?
In what tongue is the truth inscribed?

I'm certain all the dots will join;
there is no meaning without theory.
A clumsy doodle will not suffice;
it must have elegance and beauty.

I follow this path to the end; no end,
only changes in transportation
as we pass lights flickering in the dark,
secret locations along the way.

Come back again in twenty years
for an answer to the question: why?
You're right, dear sceptic, experience
alone decides what is truth or lie.

Distant parallelism? Five dimensions?
My god is mute, but leads me on.
Is there malice in his silence?
Or has he gone?

Un-American

1951–54

*"In America, even lunacy is mass-produced
and everything goes out of fashion."*

The abuse reaches him through the mist
as he walks his tightrope above the torrent,
balancing between extremes.

Who wants to be the last man standing?
Is it preferable to drown in hysteria
than swim against white water?

The spectre of a Reich stalks him
as he teeters in the cold wind,
First Amendment on his lips.

The cataract rages below, the FBI checks,
the pendulum swings, the chasm is crossed.

Playing Dice

Princeton 1948-55

"This has been a fleeting visit to a strange house. Time to go, elegantly."

Time is a stubborn illusion
but within my abdomen
an aortic clock is ticking.
I am content.

At the age of seventy
I think of Schrödinger
as I open my birthday box
to reveal a half-dead parrot.

I talk to it, tell it bad jokes,
bring it back to health.
Would it have survived
in Copenhagen?

I must seem like an ostrich,
head buried in relativistic sand,
to avoid facing the evil quanta
or Germany rearmed.

Uncertainty is everywhere.
I work in isolation, a human being
mouthing words of peace
as the nuclear world falters.

Time is a stubborn illusion
yet the aneurism persists.

Einstein's Brain

Pathology, Princeton April 1955

Deep within lies the sign of omega
where Mozart has cast his spell.
Harvey traces each convolution
for the telling imprint of excellence.

He dissects the essence of genius,
a force field that can warp the cosmos,
play mind games with Schrödinger's Cat,
subject the entire universe to new laws.

Harvey picks up a knife,
cuts down into dead matter.
It is far too late to reveal
a theory of everything.

Making Waves
Gravitational waves 1916–2016

He wasn't always sure
they could actually exist,
like marital harmony
or world government,

the phantom, perhaps,
of a fevered mind,
moving at thought-speed,
too difficult to detect

until lasers picked out
a ripple in space-time,
pinched like a pimple
between thumbs of darkness.

Somewhere between here
and the nearest star:
a single hair –
now found.

It has taken a century,
travelled a billion light years.
This strand has shape, a signal
to unravel the mysteries

of an unseen darkness,
the birth throes of a universe.

Notes

All epigraphs are Einstein's words, unless otherwise stated.

Johnnie & Dollie – This poem is from the love letters between Einstein and his wife-to-be Mileva Marić. He spent vacations with his mother and sister at the Hotel-Pension Paradise, near Zurich. The Yiddish expression *loch im kopf* means 'hole in the head'.

Thought Experiments – Lieserl Einstein (1902–?) was the first child of Einstein and Mileva, born before their marriage in 1903. She was probably put up for adoption and may have died of scarlet fever in late 1903.

Quantum Leap – Max Planck (1858–1947), a German theoretical physicist, introduced the quantum in 1900. He considered it a mathematical fudge rather than a physical reality. Einstein used the concept in 1905 to explain the photoelectric effect.

Unseen – Robert Brown (1773–1858) noted the random motion of pollen grains suspended in water but couldn't explain the phenomenon. Einstein's 1905 paper explained 'Brownian motion' as the grains being jostled through collisions with water molecules. This supported the contested theory that matter consists of discrete atoms.

Torch Bearer – Galileo argued in 1632 that motion with constant speed in a straight line can't be detected and feels the same as not moving at all. In 1905 Einstein incorporated this into Relativity Theory along with the constancy of the speed of light.

Objective Reality – Mileva Marić ('Mitsa') (1875–1948) came from Serbia to study physics at Zurich Polytechnic. She was the only woman in Einstein's class and was technically excellent but sacrificed a promising academic future to marry him. They separated in 1916 and divorced in 1919.

The New Violin - During May 1919's total solar eclipse, the British astronomers Arthur Eddington and Frank Dyson undertook expeditions to verify the prediction of Einstein's General Theory of Relativity concerning the degree of bending of light rays by the sun's gravitational pull.

When You Look at Something - The 5th Solvay Conference met in 1927 in Brussels to discuss the newly formulated Quantum Theory. Einstein demanded a reality independent of observers and not based on uncertainty. He lost the argument but never accepted defeat.

"Not Yet Hanged" - Following Hitler's rise to power in 1933, Einstein returned to Europe but realised that a return to Germany was out of the question. A German magazine noted that Einstein was "not yet hanged". He returned to America and never left.

Dilemma - Werner Heisenberg (1901–76) headed the unsuccessful German atom bomb program during WW2. The fear of a Nazi atomic bomb led Einstein to sign a letter to Roosevelt in 1939 that led to the Manhattan Project and the A-bombing of Hiroshima and Nagasaki.

No - Einstein's second wife Elsa (1876–1936) was his first cousin. They had an affair and married soon after his divorce from Mileva. Eduard Einstein ('Tete') (1910–65) was the second son of Einstein and Mileva. He suffered from schizophrenia and spent much of his life in a Swiss institution. Helene Savić (1871–1944) was a lifelong and intimate friend of Mileva.

Playing Dice - Erwin Schrödinger (1887–1961) was an Austrian physicist who joined Einstein in expressing unease at the uncertainty at the heart of quantum theory, exemplified by the so-called Copenhagen interpretation, which excludes any single underlying reality independent of an observer.

V.

Einstein's Brain - After Einstein's death in 1955, his brain was removed by Thomas Harvey, the pathologist at Princeton Hospital. Over the years, Harvey doled out pieces of the brain to researchers. As expected, no firm conclusions have been drawn concerning the correlation between Einstein's brain structure and the mind of a genius.

Making Waves - Existence of gravitational waves was predicted in 1916 by the General Theory of Relativity. They were first observed in February 2016. This stretching and squeezing of space-time originated from the merging of two massive black holes. The sensitivity required is that of comparing the width of a human hair to 25 billion miles.

Acknowledgments are due to the authors, editors and publishers of the following texts from which I have drawn material: *Einstein: His Life and Universe* by Walter Isaacson (Simon & Schuster, 2007); *'Subtle is the Lord...': The Science and the Life of Albert Einstein* by Abraham Pais (Oxford University Press, 1982); *Ideas and Opinions by Albert Einstein,* edited by Carl Seelig (Souvenir Press, 1973); *Albert Einstein/Mileva Marić: The Love Letters,* edited by Jürgen Renn & Robert Schulmann (Princeton University Press, 1992); *In Albert's Shadow: The Life and Letters of Mileva Marić,* edited by Milan Popović (John Hopkins University Press, 2003).

V.

V.

Martin Zarrop is a retired mathematician who wanted certainty but found life more interesting and fulfilling by not getting it. He started writing poetry in 2006 and can't stop. His pamphlet *No Theory of Everything* (2015) was one of the winners of the 2014 Cinnamon Press pamphlet competition. His first full collection *Moving Pictures* was published by Cinnamon Press in October 2016.

V.